Astronomy made known as never before: The james webb space

Joseph L. Smith

Copyright page

All rights reserved. No part of this publication may be reproduced, distributed, or transmitted in any form or by any means, including photocopying, recording, or other electronic or mechanical methods, without the prior written permission of the publisher, except in the case of brief quotations embodied in critical reviews and certain other noncommercial uses permitted by copyright law.
Copyright ©Joseph L. Smith, 2022.

TABLE OF CONTENT

Chapter One: The History of the James Webb Telescope.

Chapter Two: What is the James Webb Space Telescope's main goal?

Chapter Three: James Webb's impact on Astronomy

Chapter Four: Why is the James Webb Space Telescope such a big deal?

Chapter Five: To the edge of existence

Chapter One: The History of the James Webb Telescope.

The James Webb Space Telescope required 30 years and $10 billion to construct, has flown over 1.5 million kilometers from Earth, and presently, we can, at last, see the primary look at its power with an assortment of pictures.

NASA has guaranteed the most profound picture of our universe that has at any point been taken, and these first arrangements of pictures are just the most vital phase in a long occupation of growing our perspective on the universe.

Be that as it may, what could you at any point anticipate from these pictures, when and where might you at any point see them this week and what will they cover?

The primary pictures from the JWST Profound picture

In an unexpected development, one picture was delivered all alone. It came out the prior night and was disclosed by President Joe Biden.

The picture shows a far-off world bunch: SMACS 0723, seen here as it was 4.6 a long time back.

This little part of the Universe covers a fix of the sky that is generally the size of a grain of sand held at a manageable distance by somebody remaining on Earth.

The enormous dazzling white lights dissipated across the picture are stars tracked down in our world, the Milky Way. While these take up a great deal of the pictures, what is more, fascinating is the more modest dabs in the middle between. These are cosmic systems and the specks that have smeared or hauled are worlds that are a lot farther away.

The picture was accomplished by making a composite of pictures taken at various frequencies over a period that totaled 12.5 hours. The SMACS 0723 world group is huge to the point that it goes about as a gravitational focal point, in which the bunch's gravity makes light twist, as though it were going through an actual focal point.

All in all, this picture was caught by the JWST, a human-made telescope, pointing at the view made by a normally made telescope, and the mix permits us to see staggeringly far-off universes.

WASP-96 b

Exoplanet Atmosphere
Photograph by NASA/ESA/CSA/STScI
While different pictures in this first gathering were all photographs taken with the JWST, this picture rather hoped to give us knowledge about space through a diagram. A light chart uncovers key data about the barometrical structure of WASP-96b, a hot goliath exoplanet with a mass like that of Saturn.

The planet is found around 1,120 light-years from us, and the data in this diagram shows water as fumes, fogs, and proof of mists.

The discoveries of this chart show the JWST could be utilized to recognize planets that could be possibly livable.

Star demise
This one next to the other examination shows perceptions of the Southern Ring

Nebula in close infrared light, at left, and mid-infrared light, at right, from NASA's Webb Telescope.

This scene was made by a white small star - the remaining parts of a star like our Sun after it shed its external layers and quit consuming fuel, however atomic. Those external layers presently structure the launched shells up and down this view.

In the Near-Infrared Camera (NIRCam) picture, the white smaller person appears to the lower left of the brilliant, focal star, somewhat concealed by a diffraction spike. A similar star shows up - yet more brilliant, bigger, and redder - in the Mid-Infrared Instrument (MIRI) picture. This white small star is shrouded in thick layers of residue, which cause it to seem bigger.

The more splendid star in the two pictures hasn't yet shed its layers. It intently circles the dimmer white midget, assisting with dispersing what it's catapulted.

North of millennia and before it turned into a white smaller person, the star occasionally catapulted mass - the noticeable shells of material. As though on rehash, it contracted, warmed up - and afterward, unfit to push out more material, throbbed. Heavenly material was sent every which way - like a pivoting sprinkler - and gave the fixings to this uneven scene.

Today, the white diminutive person is warming up the gas in the internal districts - which seem blue on the left and red on the right. The two stars are illuminating the external districts, displayed in orange and blue, separately.

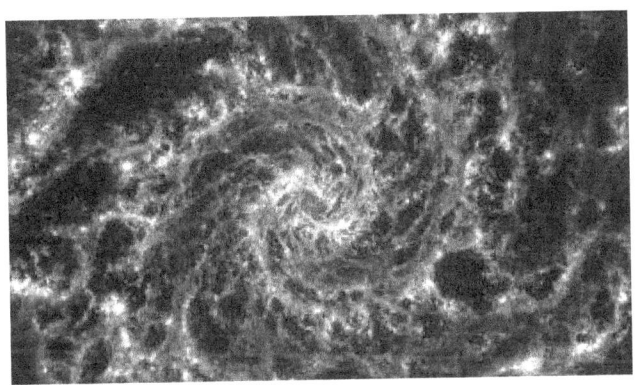

The pictures look altogether different because NIRCam and MIRI gather various frequencies of light. NIRCam sees close infrared light, which is nearer to the noticeable frequencies our eyes recognize. MIRI goes farther into the infrared, getting mid-infrared frequencies. The subsequent star all the more plainly shows up in the MIRI picture, since this instrument can see the sparkling residue around it, making it all the more visible.

These staggering pictures created by the JWST are of the Southern Ring Nebula.

At the focal point of the cloud is a withering star (found in the blue segment of the picture on the left).

This withering star is effectively ousting gas and residue which is making the orange nearly froth-like substances.

The picture on the right is of a similar Southern Ring Nebula. Not at all like the one on the left which was taken utilizing an infrared camera, this is concentrated on through the Mid-Infrared Instrument (MIRI) which assists stargazers with catching subtleties that were recently concealed in the dust.

Stephan's Quintet
A gathering of five worlds that show up near one another overhead: two in the center, one toward the main, one to the upper left, and one toward the base. Four of the five have all the earmarks of being

contacted. One is fairly isolated. In the picture, the cosmic systems are huge compared with the many more modest (more far off) universes behind the scenes. Each of the five universes has radiant white centers. Each has a marginally unique size, shape, design, and shading. Dispersed across the picture, before the cosmic systems are the number of forefront stars with diffraction spikes: dazzling white focuses, each with eight splendid lines transmitting out from the middle.

Initially distinguished as far as possible back in 1877 by Édouard Stephan, Stephan's Quintet is composed of five communicating universes that resemble practically contacting one another. While it seems as though every one of the universes is a similar distance away, one of them (NGC 7320) is much closer, a ways off of around 40 million light-years away.

Different cosmic systems are around 290 million light-years away. This gathering of universes being so close to one another permits researchers to observe their converging to get a superior comprehension of how cosmic systems develop.

While this gathering of cosmic systems has been envisioned previously, JWST accomplishes more detail than previously. Shimmering bunches of millions of stars and starbursts should be visible around beyond the quintet, and the picture likewise catches colossal shock waves as one of the systems crashes through a group.

Squint your eyes and this photo could well be a hilly reach around evening time.

This is truth be told the edge of a youthful star-shaping locale in the Carina Nebula. This picture uncovers recently darkened areas of star birth.

It has been nicknamed the 'inestimable precipices' and seeing why is clear. This is the edge of a vaporous depression that is around 7,600 light-years away. This arrangement has been made by the extreme bright radiation and heavenly breezes from supermassive, hot youthful stars situated in the focal point of the air pocket (found simply above what is perceptible in this picture.

Over the orange 'mountains' is a white steam-like substance. This is ionized gas and hot residue that is streaming away from the cloud because of extreme radiation.

These perceptions and pictures of NGC 3324 will assist with revealing insight into star developments and the introduction of stars.

When will the primary pictures emerge and where might you at any point see them?

Initially, the main pictures were all going to be delivered together, yet there has now been a shift in direction.

The principal official picture from the James Webb Space Telescope will be uncovered by Joe Biden on 11 July 2022 at 10 pm BST. This is supposed to be the most profound and most noteworthy goal of infrared perspective on the universe at any point caught.

The JWST group has now declared the rundown of articles it has focused on with its most memorable round of pictures. There will be five regions that are displayed in these pictures:

Carina Nebula

Both one of the biggest and most brilliant nebulae overhead, the Carina Nebula is approximately 7600 light-years away. The Carina Nebula houses a few gigantic stars, a few of which are a lot bigger than the Sun.

WASP-96b

WASP-96b is a goliath planet that is tracked down beyond our planetary group. It is generally made out of gas and is around 1150 light-years from Earth. It was found back in 2014.

Southern Ring Nebula

An extending haze of gas encompassed by a perishing star, the Southern Ring Nebula is the ideal chance to test the James Webb Space Telescope's infrared pictures. It is almost a portion of a light-year in measurement and is around 2000 light-years from Earth.

Stephan's Quintet

Stephan's Quintet is found 290 million light-years away. It was the primary smaller cosmic system bunch at any point found as far as possible back in 1877 and will be the uttermost picture taken by the JWST.

SMACS 0723

SMACS 0723 is a fix of the sky in the southern heavenly body of Volans. It has a monstrous group of systems in the closer view which carries on like a huge amplifying glass. This is because their staggering mass causes an observable ebb and flow of the space-time around them, amplifying the light from far-off objects.

Where is the James Webb Space Telescope now?
The James Webb Space Telescope is currently in L2 Orbit - its last objective, around 1.5 million kilometers from Earth. This is an excursion that requires approximately a month to finish.

You can keep tabs on its development with NASA's 'Where is Webb' highlight. Not in the least does this show the ongoing separation from Earth yet additionally the telescope's speed, temperature, how long it has been in a circle, and what its next stage is.

When did the James Webb Space Telescope send off?

The James Webb Space Telescope was sent off on Christmas Day 2021.

If you were too busy opening presents and partaking in a Christmas supper to watch the send-off on TV, you can see the send-off again on the JWST YouTube channel.

While the telescope has now been formally sent off, it saw an enormous number of deferrals arrive at this point. The observatory was initially expected to be sent off back in 2007. From that point forward, it has encountered north of 16 send-off delays with the pandemic broadening the date far beyond the last expected date of March 2021.

The telescope was sent off on the Ariane 5 rocket. This is a particular rocket intended to take satellites and different payloads into a move or low-Earth circle.

You may be thinking, who gets the distinction of having such a memorable telescope named after them? Indeed, that title goes to James Edwin Webb, the second overseer of NASA, most popular for heading up Apollo - the main space program to send people to the Moon.

He was additionally instrumental in the two maintained space programs that followed on from Apollo: Mercury and Gemini. While Webb did ultimately kick the bucket in 1992, matured 85, he abandoned a gigantic inheritance, meriting a telescope named after him.

"Accommodating Hubble's replacement be named to pay tribute to James Webb.

Because of his endeavors, we got our most memorable looks at the emotional scene of space," said previous NASA overseer Sean O'Keefe about the observatory's name. "He took our country on its most memorable journeys of investigation, transforming our creative mind into the real world."

The telescope hasn't forever been named after Webb. It began its life being known as the Next Generation Space Telescope which, all things considered, isn't the most innovative name we've heard!

How large is the James Webb Space Telescope?

Charged as the replacement for the Hubble Space Telescope, the JWST is the biggest space observatory at any point fabricated. Its huge sun safeguard base estimates a gigantic 22m by 12m, generally a similar size as a tennis court.

Albeit almost two times as large as Hubble (which is just 13m long), the JWST is close to around 50% of the load at 6,500kg.

The JWST's gold-plated mirrors have a complete measurement of 6.5m, a lot bigger than Hubble's 2.4m breadth plate. In general, the JWST will have roughly a 15 times more extensive view than Hubble.

How far could James Webb at any point in the Space Telescope see?
Utilizing its infra-red telescope, the JWST observatory will look at objects over 13.6 billion light-years away.

Due to the time it takes light to traverse the Universe, this implies that the JWST will be checking out at objects 13.6 quite a while back, an expected 100 to 250 million years after the Big Bang. This is the farthest back in time at any point seen by mankind.

Where will the James Webb Space Telescope circle?

After sending off into space, the JWST will circle the Sun, flying up to 1.5 million kilometers from Earth in temperatures coming to - 223°C.

For correlation, the Moon is 384,400km away, while the Hubble Space Telescope flies just 570km over our planet. As the JWST will work so distant from Earth, it can not be overhauled by space travelers if any flaws emerge.

Chapter Two: What is the James Webb Space Telescope's main goal?

As the JWST is a result of a global joint effort between NASA, the European Space Agency (ESA), and the Canadian Space Agency (CSA), it has numerous mission objectives.

These include:

Analyze the primary light in the Universe and the divine items which framed not long after the Big Bang.
Research how worlds structure and develop.

Concentrate on the airs of far-off exoplanets.

Catch pictures of planets in our planetary group.

Find proof of the dim matter.

The JWST is supposed to work for a long time after its send-off. Be that as it may, NASA trusts the observatory will endure longer than 10 years.

Tragically, the observatory will not have the option to work until the end of time: albeit general sunlight based controlled, the JWST needs a limited quantity of limited fuel to keep up with its circle and instruments.

How is the James Webb Space Telescope different from Hubble?

The James Webb is seen in numerous ways as a better replacement for the Hubble telescope which was sent off way back in 1990. Be that as it may, would they say they are comparable, or are these two telescopes radically unique?

The two telescopes, right off the bat, see light in various ways. The Hubble's principle center is around both noticeable and bright light. While it can notice an exceptionally little piece of the infrared range, it is not even close to the degree to which the JWST can.

The JWST is explicitly intended to zero in on the infrared range. It can't find in that frame of mind as Hubble would be able, yet it will want to zero in on brilliant articles like extremely far-off universes.

The James Webb Telescope is likewise a lot bigger than the Hubble, for the most part, because of its huge sunshield. This is utilized on all space telescopes yet is particularly significant with the James

Webb because of its infrared cameras. If it isn't kept cool, it could gamble by blinding itself to the lights of articles it is attempting to notice.

Another critical distinction between the two satellites is the distance that they will be kept. The Hubble telescope circled over Earth's climate yet was close to the point of being drawn nearer on the off chance that fixes should have been finished.

The JWST then again will be far away, around 1.5 million kilometers away! That is both farther than any human has at any point voyaged and excessively far for anybody to at any point go fix the satellite assuming something turns out badly.

It will be this far out for a couple of reasons. It will be where the gravity of the Sun and Earth cooperate to assist with keeping the satellite set up, in addition to it will be far away from the reflected radiation of the Earth, assisting keep it cooling.

The reality of the James Webb Telescope.

10 amazing however confirmed realities about NASA's James Webb Space Telescope.

On December 25, 2021, accepting an unanticipated inconvenience, the James Webb Space Telescope will send off from French Guiana. While stargazers hold their aggregate breaths, trusting that each vital step will go just before science tasks start, we as a whole can all in all value what a wonder the telescope is. The following are 10 realities — random data to some, the outcome of a vocation of difficult work for other people — for everybody to enjoy. The most postponed telescope in history is going to encounter not simply a decision time, but rather a progression of them over the approaching not many months. To begin with, the telescope should endure its December 25 send-off, which should point it definitively on course to the L2 Lagrange point. Then, at that point, it should effectively isolate itself from the send-off vehicle and afterward convey its sunlight-powered chargers very quickly. From that point forward, the pinnacle gathering, the sunshield, and the essential and auxiliary mirrors should all effectively send steps including many weak link components. A progression of engine firings should likewise occur, in the long run prompting Webb to show up at its objective: in a circle around the L2 Lagrange point. If — and provided that — these means succeed, then NASA's James Webb Space Telescope won't start accepting information as ever previously, investigating the Universe with remarkable power and an unparalleled series of instruments and capacities. There is a progression of disclosures we're essentially ensured to make once science tasks start, as well as the potential for finding whatever dwells out there amid the tremendous expanse of the obscure cosmos. And yet, regardless of that, it's all likewise worth valuing a portion of the astonishing and novel design that is gone into the plan and execution of this telescope. Right away, the following are 10 unimaginable and difficult-to-trust realities about NASA's best-in-class observatory: the James Webb Space Telescope.

1.) The James Webb Space Telescope is lighter than its ancestor, the Hubble Space Telescope. This one is a genuine stunner to a great many people.

Under most conditions, if you need to construct a greater rendition of something, it will be heavier and more gigantic. For comparison: Hubble was 2.4 meters in width, with a strong essential mirror, and a gathering area of 4.0 square meters.

James Webb is 6.5 meters in width, made from 18 different mirror fragments, with a gathering area of 25.37 square meters.

But, if we somehow happened to put them both on a scale here on Earth, we'd find that Webb has a mass of ~6,500 kg or a load of 14,300 pounds. At the point when Hubble was sent off, for examination, it had a mass of ~11,100 kg and a load of 24,500 pounds; with its updated instruments, it currently has a mass of ~12,200 kg and a load of 27,000 pounds. This is a huge accomplishment of designing, as basically every part on

James Webb, where material is lighter than its Hubble simple.

2.) James Webb's mirrors are the lightest huge telescope reflections ever. Every one of the 18 essential mirror sections, when it was previously made, looked like a bent plate, and had a mass of 250 kg (551 pounds). By the time they're finished, in any case, that mass has been decreased to a simple 21 kg (44 pounds), or a 92% decrease in weight. The way this is achieved is entrancing. To start with, the mirrors are cut into their hexagonal shape, which offers a slight decrease in mass. However at that point — and here's where it gets splendid — essentially all of the mass on the "rear" of the mirror is machined away.

What remains has been tried to guarantee that it will:

hold its exact shape considerably under the burdens of send-off

not break under vibrations and strain, regardless of its fragile nature

endure the normal number and speed of micrometeoroid influences

be delicate to the required changes in shape that will be changed by the actuators joined to the back, on the whole, these 18 mirrors will frame a solitary mirror-like plane to a precision of 18 to 20 nanometers: the best, all with the lightest such mirrors at any point made.

3.) Although they seem gold, James Webb's mirrors are made from beryllium. Indeed, there's a gold covering applied to every one of the mirrors, however, it would have been disastrous to produce the mirrors completely out of gold. Actually no, not in light of the exceptionally high thickness, nor on account of gold's flexibility, the two of which are properties it certainly has. The enormous issue would be warm expansion. Even at exceptionally low temperatures, gold extends and contracts considerably with little temperature changes, which is a dealbreaker for the material of decision for Webb's mirrors. Be that as it may, beryllium radiates on this front. By chilling off beryllium to cryogenic temperatures and cleaning it there, you guarantee that there will be room-temperature defects, yet that those flaws will vanish when those mirrors are cooled in the future to working temperatures.

Just once the beryllium is produced and machined to its last shape is the gold covering then, at that point, applied.

4.) The aggregate sum of gold in the James Webb Space Telescope's mirrors is just 48 grams: under 2 ounces. Every last one of James Webb's 18 mirrors should be extraordinary at mirroring the sort of light it's intended to notice: infrared light. How much gold applied should be perfect; apply close to nothing and you won't cover the mirror altogether, yet apply excessively and you'll begin to encounter development, compression, and deformity when the temperatures change. The process by which the gold covering is applied is known as the vacuum fume statement. By setting the "clear" mirrors inside a vacuum chamber, where everything the air is emptied, you then infuse a limited quantity of gold fume inside. Regions that needn't bother with being covered, similar to the rear of the mirror, are veiled off, so just the smooth, cleaned surface breezes up covered with gold. This cycle goes on until the gold arrives at the ideal thickness of just ~100 nanometers, or about ~600 gold molecules thick.

By and large, there are just 48 grams of gold in the James Webb Space Telescope's mirrors, while the dull posteriors get swaggers, actuators, and flexors joined to them.

5.) The actual gold won't be straightforwardly presented to space; it's covered in a slender layer of undefined silicon dioxide glass. Is there any valid reason why you wouldn't simply uncover the actual gold in the profundities of the room? Since it's so delicate and pliable, it's exceptionally powerless to harm from even a gentle or minuscule effect. While the beryllium is generally unaffected by micrometeoroid influences, a meager gold covering would be, and would thus not be able to keep up with the perfection fundamental for the telescope's activity without an extra layer of security.

That is where the last "covering on the covering" comes in: undefined silicon dioxide glass. Even though we ordinarily partner mirrors with being made from glass with some kind of covering on it, the capability of the glass is extremely basic for this situation: to be straightforward to the light and to safeguard the gold. So indeed, it's gold covered, however at that point, the actual gold should be safeguarded with its covering as well.6.) The "telescope side" of James Webb will latently cool itself down to no higher than ~50 K: adequately cool to make nitrogen liquify. The entire explanation James Webb should be put so distantly from Earth, at the L2 Lagrange point rather than in a low-Earth circle like Hubble, is because it will be latently cooled as at no other time. A colossal five-layered sunshield has been uniquely made for James Webb, reflecting however much of the daylight away as could be expected and safeguarding the layer underneath it. On the off chance that it was in the low-Earth circle, the infrared intensity produced by the Earth would keep it from arriving at the fundamental low temperatures. The precious stone-formed sunshield itself is huge: 21.2 meters (69.5 feet) in the long aspect and 14.2 meters (46.5 feet) in the short aspect. Each layer has a "sweltering side" that points toward the Sun and a "cool side" that faces the telescope. The peripheral layer will, on its hot side, arrive at a temperature of 383 K, or 231 °F. When you get to the deepest layer, the hot side is just 221 K, or - 80 °F, however, the virus side is right down to 36 K, or - 394 °F. Insofar as the telescope stays underneath ~50 K, it will be fit for working as designed.7.) With dynamic, cryogenic cooling, Webb will get right down to ~7 K. All the low temperatures arrived at by uninvolved cooling, in the 36-to-50 K reach, are adequate for the activity of Webb's close to infrared instruments. This incorporates three of its four significant science instruments: NIRCam (the close-to-the infrared camera), NIRSpec (the close-to infrared spectrograph), and the FGS/NIRISS (fine-direction sensor/close-to infrared imager and slitless spectrograph). They're undeniably intended for activity at 39 K: well inside the scope of uninvolved cooling. But the fourth instrument, MIRI (the mid-infrared imager), should be cooled significantly farther than detached cooling can get you, and that is where the cryocooler comes in. Helium just becomes fluid at around 4 K, thus by connecting a fluid helium fridge to the MIRI instrument, Webb researchers can chill it off to the required working temperature: ~7 K. The more drawn out the frequency of light that you need to test, the cooler you want to get your instruments, which is the essential justification behind a large portion of the plan choices that went into the James Webb Space Telescope.

8.) Unlike NASA's Spitzer, which progressed to a "warm" mission when it ran out of coolant, James Webb ought to keep up with its chilly temperatures for its whole life expectancy. The fluid helium that keeps James Webb effectively cooled, on a basic level, ought to never run out; it's a shut framework. In any case, as anybody who's consistently worked in trial physical science can authenticate, spills occur, regardless of how well you defend against them. Intended for a 5.5-year mission, at least, with the chance of 10 years or longer under the most hopeful conditions, Webb shouldn't run out of its cryogenic coolant assuming it satisfies its plan specifications however, there's generally the likelihood that something will turn out badly, and we will not have the option to effectively cool the mid-infrared imager adequately or for the whole mission, and that will eat into

Webb's responsive qualities at dynamically increasingly long frequencies. (A similar provision applies to the close to infrared instruments in case of sunshield harm or shortcomings.) The hotter the James Webb Space Telescope gets, the smaller its frequency range it can test will become.9.) When it runs out of fuel, its destiny will be to for all time dwell in a "burial ground circle" around the Sun. Hubble, with help from four overhauling missions, is as yet working more than three entire long times after its send-off. Webb, nonetheless, necessitates to utilize its fuel at whatever point it maintains that should do anything including movement. That incorporates:

to play out a consume to address its course towards its objective at L2
to perform orbital rectifications to keep it in its circle at L2

to situate itself with the goal that it focuses on its ideal targetFuel arrives in limited stock, and the amount we have left for science tasks relies completely upon how much the send-off puts Webb in its optimal direction toward its final location.

Chapter Three: James Webb's impact on Astronomy

Everything you wanted to know about upcoming missions to learn about the cosmos.

Webb's 18 mirror segments can lock into the largest telescopic mirror humankind has ever built.

The Five Big Ways the James Webb Telescope Will Help Astronomers Understand the Universe.

The James Webb Space Telescope, the much anticipated, revolutionary newcomer to the pantheon of space telescopes, is finally set to make a big splash—a welcome change from the waves its oft-delayed launch and ballooning cost have made. After several hold ups this year, NASA is set to launch the craft on Christmas. The telescope's takeoff has now been delayed a decade, and its cost has risen roughly $9 billion over budget. Lawmakers and scientists both have expressed concern that the project is siphoning funds from other research areas, but many other scientists say that Webb is worth the money and the wait.

Webb's conception is inspired by the Hubble Space Telescope—the 31-year-old observatory famous for capturing stunning photos of our universe's galaxies. But Webb picks up where its predecessor falls short, says Eric Smith, Webb's program scientist and chief scientist of NASA's Astrophysics Division. There's no telescope like Webb so far, he says. The new observatory, which is slated to launch from northern French Guiana near the equator, is a collaboration between the space agencies of the United States, Europe, and Canada. "When you see Webb go into space, ... it's the whole force of human creativity and all kinds of disciplines that push it there."

This latest space-bound contraption is unique because of two capabilities. First, it's big, with a 21.3-foot primary mirror that will make Webb the farthest-seeing telescope humankind has ever built. Secondly, Webb views the universe in the infrared—the zone on the electromagnetic spectrum with slightly longer wavelengths than visible light. It will be the only infrared-specialized telescope in space that can see long distances. Its closest challenger, Hubble, works primarily in the visible and has a limited infrared-viewing range.

"Anytime astronomers get a new telescope, it's a kid in a candy store kind of thing," says Smith.

Partially thanks to Webb's size and infrared gaze, here are five things the telescope will allow astronomers to do.

Understand how early galaxies formed and grew
Rubin Galaxy

The Rubin Galaxy, named after the astronomer Vera Rubin, twirls in space 232 million light years away from Earth. NASA, ESA, and B. Holwerda, University of Louisville

"One of the big purposes of telescopes is actually as time machines because the distance is look-back time," says Daniel Eisenstein, an astrophysicist at the Harvard–Smithsonian Center for Astrophysics. Eisenstein will use Webb's cameras to "time-travel" back to when the earliest galaxies were forming right after the Big Bang.

When we look at a distant galaxy light years away, we aren't seeing it in its most recent state. Its distance in light years translates to the number of years it takes for its light to arrive on Earth. For example, the closest galaxy to ours is the Canis Major Dwarf Galaxy which is 25,000 light-years away, so its light takes

25,000 years to reach Earth. That means when we look at Canis Major Dwarf, we're seeing it as it was 25,000 years ago.

The further into space scientists can look, the further back in time they can observe a galaxy. Webb, being the farthest-seeing telescope yet, can root out the youngest-looking galaxies humanity can observe. To understand the formation of galaxies, scientists like Eisenstein will look at several galaxies at different life stages and piece together their developmental timelines.

Webb's infrared capabilities are also crucial for observing these galaxies. Light from distant galaxies will be stretched out by the expanding universe. By the time the light reaches our telescopes, its original wavelength will have shifted from the visible or ultraviolet to the infrared. Luckily, picking up infrared signals is right up Webb's alley. "It's the first time we've had a large, cold telescope in space that can observe these infrared wavelengths," says Eisenstein.

The Hubble space telescope has managed to capture the shortest wavelength infrared rays stretched from the bluest of light of faraway galaxies. The retired Spitzer infrared telescope was much smaller than Webb and couldn't see as far into space. Webb will knock it out of the park in terms of how deep into space—and how far back in time—it can catch distant galaxies in the act of growing up.

Detect possible chemical signatures of life on other planets
Exoplanet from moon
An artist's impression of the view of an exoplanet from its moon. The James Webb Space Telescope will allow

scientists to look for signs of life on such planets. IAU / L. Calçada

If life exists outside of Earth, it will release distinct chemical signatures, such as breathing carbon dioxide and photosynthesizing out oxygen, that can transform a planet. Analyzing the chemicals in a planet's atmosphere will not only allow scientists to look for life but also enable them to assess a planet's habitability.

Webb can detect infrared wavelengths for fingerprinting chemicals such as water and methane present in the atmosphere of exoplanets, which are planets beyond our solar system.

Webb contains two instruments that will allow scientists to unravel the wavelengths of infrared signals from solar systems beyond ours—to unweave the colors of the infrared rainbow, so to speak. When an exoplanet photobombs a star that our telescopes are gazing at, the starlight will experience a dip in certain energies corresponding to the chemicals in the exoplanet's atmosphere. If Webb happens to be looking at the right star at the right time, it can chemically analyze the atmosphere of the star's planet by analyzing the blip in the starlight.

"Exoplanet science as a field is pretty new," says Munazza Alam, an astrophysicist at Harvard–Smithsonian Center for Astrophysics. Since the first exoplanet discovery in 1992, scientists have found thousands of exotic planets teeming in the universe. "They're everywhere," she says.

However, humanity's understanding of these exoplanets has barely extended beyond the fact that they're there. It's challenging for current technology, such as Hubble or on-Earth infrared telescopes to carry out infrared spectroscopy on new exoplanets of interest. Hubble works with a much narrower band of infrared energies compared to Webb. Ground observatories are shrouded in Earth's atmosphere, which itself is an absorber and scatterer of infrared light. The Earth also emits background infrared radiation that would overwhelm the faint signals coming from the deep cosmos. In space where Webb will be, Earth's atmosphere and warm radiating surface are out of the way for an unobstructed view of the night sky.

Alam will be using Webb to survey a motley of Jupiter and Neptune-sized exoplanets, and she's excited to get started. "We're just at the tip of the iceberg."

Watch the birth of stars

Visible IR

The "Pillars of Creation" in the Eagle Nebula look starkly different when seen under visible light (left) and infrared (right) by Hubble. Webb's infrared capabilities will allow scientists to peek beyond the dusty veil of stellar nurseries to capture snapshots of star formation. NASA, ESA / Hubble, and the Hubble Heritage Team

The birthplaces of stars are full of dust. While they make for breathtaking photos, the dust blocks scientists from peering right into the heart of these clouds when they look at them with visible light. Luckily, infrared light from stars can penetrate through the dust to give scientists a whole new take on an old view.

"Red light can pass through the dust in the Earth's atmosphere better than the shorter wavelengths, the blue lights," says Marcia Rieke, an infrared astronomer at the University of Arizona who is the principal investigator of one of Webb's infrared cameras. The same principle explains why infrared light can penetrate even further through dusty galaxies than visible light. "If you look at the setting sun, it tends to look much redder than when you look at it in the daytime; it's the same thing."

Hubble's limited infrared capabilities have barely scratched the surface for studying stellar formation; Webb's broader infrared range will enable scientists to peer deeper into the dust.

Young stars emerge from the dustiest pockets where it's most challenging to see through. Thanks to Webb's high infrared sensitivity and spectacular resolution, scientists might be able to sift through the dust to make out these infant stars with unprecedented detail. And Webb might help scientists figure out how the dust cooks up a star, why stars form in clusters, and how planets form around a star.

Study black holes from a different angle
Black hole
A supermassive black hole such as the one at the heart of the Messier 87 galaxy is visible thanks to the stellar matter that surrounds it. Webb will help scientists observe the stars orbiting the black hole. EHT Collaboration via Flickr
Nothing can escape a black hole, not even light; so technically, black holes are invisible. Luckily, swirling around black holes is plenty of stuff we can see—stars, dust, and entire galaxies. To study black holes, scientists scrutinize this stellar menagerie, similar to studying a shadow to learn about its shadow-casting object.

In the past, scientists have used X-ray telescopes to study specific kinds of physics of black holes. These telescopes look at phenomena that are millions of degrees hot and high enough to produce X-rays, such as the violent shredding of stars wandering too close to a black hole. Webb's infrared instruments will allow scientists to see different goings-on in a black hole's corner, particularly, the cooler gasses and stars dancing around their invisible neighbor.

Where stars congregate is a dusty place; luckily, Webb's infrared eye will allow scientists to peer past the dusty curtain and see through it all. Webb will provide valuable data to peek into the temperatures, speeds, and chemical compositions of the stellar cloaks of black holes. Scientists can use this data to learn more about the mass and size of the black hole, and more about how it snacks on a star.

Be surprised by unexpected discoveries
Webb in space
An artist reimagines Webb and its sunshield in space. Northrop Grumman
Webb will be the first telescope of its kind in terms of its size, sensitivity, and wavelength range altogether. With its capabilities, a good chance exists scientists will get to see something they've never seen before.

"There's, of course, what we don't expect, and that'll probably be the most exciting," says Rieke. Maybe a phenomenon that completely overturns existing theories about the universe, she suggests.

Like many scientists crossing their fingers for Webb's successful launch, she's impatient for Webb to leave. "I want it to get launched and I want to start getting the data because I'm not getting any younger." Rieke has been involved in Webb's conception since 2001 and was also a co-investigator of Hubble's infrared instruments from the 1980s to the early 2000s. "And I'd like to have all this done before I decide it's time to retire."

She had expected to retire five years ago, but Webb's launch delays convinced her to delay her retirement. Now with Webb's launch seemingly inevitable, she hopes she can complete in four years a study on the universe's very first galaxies, after which she will finally take her long overdue break. But, she says, if she finds something entirely new and mind-boggling, she might just be willing to put off retirement for a little bit longer.

Chapter Four: Why is the James Webb Space Telescope such a big deal?

The James Webb Telescope launched into a gray sky, its blazing trail reflected in the waters of a lake.
Arianespace's Ariane 5 rocket launches with NASA's James Webb Space Telescope on board, from the ELA-3 Launch Zone of Europe's Spaceport at the Guiana Space Centre at Europe's Spaceport, at Guiana Space Center on December 25, 2021, in Kourou, French Guiana.

Perseverance Mars rover damaged by pebble flung in gust but functioning fine

Here comes James Webb Space Telescope's first full-color photo drop

New sailplanes to "fly for free" and collect data over Mars

NASA's Perseverance rover sends back Mars soundscape playlist

NASA fires its first Australian rocket launch in 27 years from Arnhem land

The new James Webb Space Telescope passed another major milestone over the weekend, deploying its primary mirror – the crown jewel of this long-awaited observatory.

The mirror is a whopping 6.5 meters in diameter, bigger than any mirror previously launched into space. (The Hubble Space Telescope's primary mirror is a mere 2.4m across.) The size heightens the sensitivity of the telescope – the larger the mirror area collecting light, the more details it can capture of a star or galaxy.

The mirror is made up of strong, light hexagonal segments tessellated together. For launch, they were folded up into two "wings" to allow for the telescope to fit into the launch vehicle. But now they have successfully stretched out again.

This follows the recent deployment of the secondary mirror: a small, circular mirror that plays a vital role in reflecting light from the primary mirror to the instruments.

Both of these mirrors are covered in a microscopically thin layer of gold. This isn't just to look fancy – it optimizes them for reflecting infrared light.

James Webb Space Telescope's primary mirror at NASA Goddard. The secondary mirror is the round mirror located at the end of the long booms, which are folded into their launch configuration. Webb's mirrors are covered in a microscopically thin layer of gold, which optimizes them for reflecting infrared light, which is the primary wavelength of light this telescope will observe.

The most exciting thing about the James Webb Space Telescope (JWST) is its promise to revolutionize infrared astronomy. With a massive mirror and the ability to see light at the infrared part of the spectrum, it can peer back billions of years through history to capture the faint, red-shifted light from the very beginning of the universe.

It will be able to watch the first stars and galaxies flicker on, probing the mysterious processes that took the universe from its dark ages and thrust us into the era of light.

Astronomers have had burning questions about this early era of the universe for decades – for example, what were those first stars like? How did magnetism and turbulence play a role in triggering the first stars to be born? How did black holes first form, start to grow, and become the hearts of galaxies?

JWST has been purposefully designed to answer these questions and more.

"Through deep observations, James Webb will reveal the very first galaxies formed in the infant universe and how those galaxies evolved across 13 billion years of cosmic time," explains Lisa Kewley, director of ASTRO 3D from the Australian National University.

"We will obtain an unprecedented picture of how galaxies like our Milky Way formed and evolved. We will measure how the elements responsible for life: oxygen, carbon, and nitrogen, formed and evolved across 13 billion years of cosmic time. James Webb will also reveal what elements are in the atmospheres around extrasolar planets.

"The big questions that James Webb aims to answer are all about our origins and our place in the universe: Are we unique? Is our Earth unique? Is the Milky Way unique? What are our origins?"

The very first chapter in the universe's history has previously been hard to study because the only way we can learn about it is through light. As the universe expands, the light of these first stars has been stretched as it travels toward us, shifting it from more energetic waves of ultraviolet or visible light into the red end of the electromagnetic spectrum.

"We cannot see a broken bone without an X-ray machine," explains Nicha Leethochawalit, an ASTRO 3D Fellow at the University of Melbourne. "Similarly, we need near-infrared wavelengths to detect galaxies at the beginning of cosmic time, and mid-infrared wavelengths to determine their compositions."

Unfortunately for us, this infrared light is the same thing as heat, and we have plenty of that on Earth, drowning out the fainter signals of galaxies far, far away.

But by launching a massive infrared-sensitive telescope into the freezing expanses of space, we can capture the flickers of these ancient stars.

"In the visible to mid-infrared wavelength range, JWST's diameter is unprecedented," Leethochawalit says. "It will detect the faintest and furthest source known to mankind."

The Hubble Space Telescope has been orbiting the planet for more than 30 years, but since it's not optimized to study the universe in infrared, it hasn't been able to answer all of the questions that JWST will tackle. Results from Hubble inspired the design of the JWST.

Hubble discovered that stars, galaxies, and supermassive black holes were around far earlier in the universe's history than we'd thought, and have evolved dramatically through time. We've also learned about the role that dark matter and dark energy play in the universe's evolution, and we've discovered thousands of exoplanets orbiting stars. JWST will tell us more about each of these things.

The bigger mirror of JWST – plus its specific infrared capability and the greater distance from the heat of Earth – means that this new telescope will take infrared astronomy above and beyond what Hubble was able to do.

So what's next on the journey?
Unlike Hubble, JWST won't be orbiting around the Earth – in fact, it's still en route to its final workplace, the second Lagrange point (L2).

Here, at this particular point about 1.5 million kilometers from Earth, the telescope will orbit the Sun – but while always staying in line with the Earth. It will keep up with our planet as it orbits, held in place by the gravity of both Earth and the Sun, so it will need relatively little rocket fuel.

James Webb Space Telescope orbit as seen from above the Sun's north pole and as seen from Earth's perspective.

This also makes it easy for its sunshield – which was also successfully deployed last week – to always protect the delicate mirror and instruments from the heat from both the Earth and Sun. The telescope has to operate at about -225°C, so this protective configuration is essential.

It will take a total of 30 days for JWST to complete its journey to L2, so there are still a couple of weeks to go before it is inserted into its orbit.

While this is happening, the mirrors are continuing to prepare themselves. They have already begun the necessary cooldown process in the shade of the sunshield, but it will take several weeks for them to reach stable operating temperatures. It's a finely controlled process, with electric heater strips managing the cooling rate so everything shrinks carefully.

As everything cools down, each mirror segment will be carefully moved out of launch configuration and into position.

All the pieces need to be perfectly aligned to correctly focus on far-off galaxies. This job is done by tiny mechanical motors called actuators attached to the back of each mirror piece, which can shift the pieces until they together act as one large coherent piece, producing sharp images.

"Aligning the primary mirror segments as though they are a single large mirror means each mirror is aligned to 1/10,000th the thickness of a human hair," explains Lee Feinberg, Webb Optical Telescope Element Manager at NASA's Goddard Space Flight Center, in Maryland. "What's even more amazing is that the engineers and scientists working on the Webb telescope had to invent how to do this."

Then, once JWST is in orbit and the mirror is ready, the scientific instruments will come online.

But it'll still take a full extra five months before JWST turns its eye to the sky for the very first time, as it will take time for the optics systems to be properly aligned and the instruments to be meticulously calibrated.

"JWST is a remarkable feat of engineering, one of the most complex instruments ever built, and will demonstrate our capacity to operate new space technologies at more than 1.5 million km from the Earth," says Simon Driver from the International Center for Radio Astronomy Research (ICRAR) and the University of Western Australia.

"Australia will play a critical role in tracking the Ariane 5 launch vehicle as it enters space, and later the routine downloading of the science data eight hours a day from NASA's deep-space tracking station at Tidbinbilla, ACT."

Looking to the future
Astronomers around the world are waiting with bated breath for JWST to begin its science program in mid-2022.

"It's been like waiting for Christmas for 20 years," says Daniel from Monash University. "The first round of observations will be particularly special for our group at Monash as we have time awarded to image a baby planet we discovered in 2018. We're hoping that a picture of an infant planet orbiting a young sun will prove one of the spectacular early results from James Webb."

But while some people have been waiting for decades, for younger scientists this will mark the beginning of exciting careers.

"During the past few years as a young researcher, I have been learning about many of the current questions and challenges in astrophysics," says Juan Manuel Espejo, Ph.D. candidate at the Center for Astrophysics Supercomputing at the Swinburne University of Technology.

"What I have found as a common theme in almost every one of them is that JWST will play a crucial role in providing answers and alleviating our curiosity for understanding the cosmos."

Claudio Lagos Urbina, from ASTRO 3D, adds: "I am so looking forward to this – as a computational astrophysicist, we have been making predictions for what JWST will see for some years and it is now the time to test these predictions, and hopefully turn those new observations into a transformational understanding of how galaxies form, and especially, how our home, the Milky Way, came to be.

"We are truly expecting a breakthrough in our capabilities of understanding the formation of the Universe."

Chapter Five : To the edge of existence

Huge telescopes can look so profound into the Universe that they can likewise think back billions of years in time. In 2018, the replacement of the Hubble Space Telescope, the James Webb Space Telescope, will want to see the period soon after the Big Bang, when the principal stars and worlds were shaped. Space experts from Leiden are helping construct instruments for the James Webb, and can hardly stand by to involve it for perceptions.

The huge number of million years back in time

If we 'just' look at 100 million light a long time around us, we see a quiet piece of the Universe, with perfect old cosmic systems that seem to be the Milky Way, the world that we are essential for. The middle between is pretty much vacant. In any case, on the off chance that you utilize current telescopes to look billions of light years away - and in this manner billions of years back in time - the 'early' universe looks altogether different for sure. Numerous worlds are turbulent in shape and produce new stars at a quick rate or have a colossal dark opening in their middle that conveys bunches of particles into space, causing shock waves there. Furthermore, albeit youngster systems can become progressively as indicated by virtual experiences, by consuming their younger siblings, they seem, by all accounts, to be huge and monstrous shockingly not long after the Big Bang.

Since its send-off in 1990, the Hubble telescope has been circling Earth. Since the telescope is over our planet's climate, it can make incredibly sharp pictures.
On to new significant distance records
Marijn Franx has been associated with the Hubble Space Telescope's journey of revelation into the early Universe all along: 'Hubble drastically changed this field of exploration at that point. In 1996, it demonstrated simply that Hubble might see such very far-off worlds since they were a lot more splendid than anticipated.' In 2016, Franx was in a group that found the ongoing significant distance record holder, system GN-z11, at 13.4 billion light years away, when the universe was just 400 million years of age.

Franx: 'Hubble has now given its best, however its replacement, the James Webb Space Telescope, will want to see a lot further back in time. We have standard conversations with individuals at NASA who are building it about the choices that are made. In such activities, you never have sufficient cash to do everything, except we as clients tell them: 'This is the very thing we truly need.'

It won't be imaginable to perform support on the James Webb, since it will be stopped in space far from Earth. So choices should have been made quite a while back about which perception instruments will be sent up with it for the following decade, and this will extraordinarily figure out what the telescope will want to see.

World GN-z11, to date the uttermost universe to be noticed, is essential for the notable Great Bear group of stars. Source: NASA

Tempestuous adolescence

Distance records in themselves are not unreasonably significant. What space experts need to comprehend is the fierce adolescence of the Universe. In cosmic systems, new stars continually structure from gas and residue mists, however as of now, on our side of the Universe, stars structure at a sluggish rate since there is not any gas and residue left between the stars. If you apply this rationale to youthful systems, they should thus have contained a ton of gas and residue. The super telescope on Earth, ALMA, was extraordinarily worked to notice the millimeter radiation that this residue emanates. What's more, ALMA is as yet mentioning observable facts in abundance.

Franx: 'We thought at the time that when James Webb was sent off ALMA would have mentioned these observable facts, yet ALMA sees not very many incredibly far-off worlds. It's a genuine secret. Maybe they contain too little residue? We have subsequently changed the James Webb program, to improve it at identifying those universes.'